D0854178

DANGER!
DO NOT OPEN

DANGEROUS EXPERIMENTS

FOR AFTER DINNER

Concept by Angus Hyland
Illustrated by Dave Hopkins
Text by Kendra Wilson

Laurence King Publishing

TABLE of CONTENTS

THE SOLE OBJECTIVE 7

LIFTING THE SPIRITS 8

THE GUILLOTINED PUDDING 11

TRIVET OF TRIUMPH 12

TABLE TURN TOSS 15

LEVER JACK SODA JERK 16

THE SOMMELIER AND THE SABRE 18

SPINNING HALOES 21

THE BENIGN FLAME 22

LA TABLE VIVANTE 25

EGG ON A TIGHTROPE 26

ORDEAL OF THE SUGAR CUBE 29

CANDLE QUADRILLE 30

WINE REVOLUTION 33

TEACHING A BOTTLE TO SUCK EGGS 34

EQUILIBRIUM OF A TOOTHPICK 37

THE SCOTCH HOP 38

PIN MONEY 41

STRONGMAN SWIZZLE STICKS 42

BATTERY OF BUBBLES 45

CORK AND FORK PIN SPIN 46

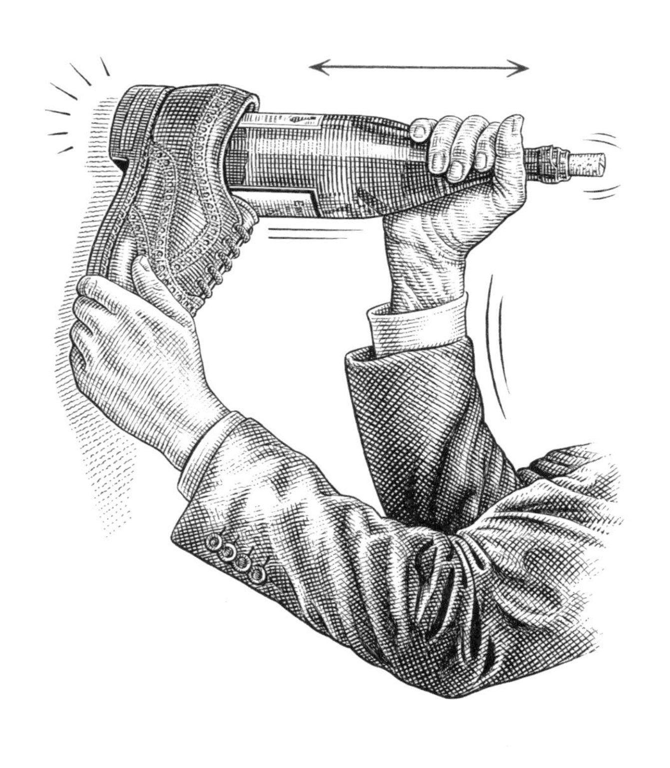

IN WHICH THE HOST
DISPENSES WITH THE CORKSCREW

THE

SOLE OBJECTIVE

EQUIPMENT

Bottle of wine with cork, comfortable shoe, thick wall for hitting

HOW INFURIATING TO arrive at a dinner party with a well-considered bottle of wine only to find that the hosts have lost their corkscrew. And yet this does happen, especially when the hosts have done it on purpose. A note to the host: By getting your friends to try this trick on arrival, you will open their minds to the possibilities of further experiments during the course of the evening. Additionally, they will take home a useful skill.

Start by removing the foil around the cork. A good-quality cork is long and made in one piece; it is more stubborn than a reconstituted cork or a synthetic one. This is just one reason an expensive bottle of wine may not be the best choice here. Next, choose a shoe that improves with wear, such as a brogue. Avoid running shoes (designed to absorb shock), ballet pumps, stilettos.

Insert the base of the bottle into the shoe, at the heel end, and hold the two so that the shoe is vertical and the bottle horizontal. Adopt a wine-opening stance: feet shoulder-width apart, knees slightly bent, arms relaxed. Note that concentration (and balance) could be impaired if one is still wearing the other shoe, or if several bottles have already been opened using this method.

Novices might begin with a white wine. Pound the cork out by striking the heel of the shoe against the wall in a steady and regular motion. After around six hits there will be a clear progression of the cork, though the speed of its exit will depend on the quality of the thrusts – and the shoe. When the cork is more than halfway out, return the bottle to a semi-upright position and ease its passage manually.

(and other drinks)

EQUIPMENT

Viscid hand, claret glass, good ordinary claret

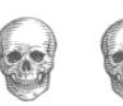
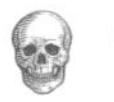

LIKE STRETCHING BEFORE exercise, a simple challenge like this will whet the appetite for more complicated experiments. It is a technique that can be demonstrated by a willing child and a glass of pure water, graduating, with the child's departure, to adults and glasses containing whisky, port or good ordinary claret.

Glasses that are suitable depend on the hands that are available: the optimum combination would be a sturdy claret glass (though not too thick) and a viscid hand – moist but not wet, tacky but not glutinous. A port glass would suit a smaller hand, a wider whisky tumbler a bigger hand. A champagne flute would suit somebody who doesn't mind breaking their long-stemmed glasses, which may already be in short supply.

The essence of this trick is to create a vacuum. The hand needs to be placed over the top of the glass with fingers pointed downwards, at right angles to the palm. Apply pressure to create a tight seal around the rim. Quickly lift the fingers so that the hand is extended flat over the glass. As the hand stretches out, the glass is suctioned on to the hand and can be lifted at the same time. It is simply a case of air pressure squaring up to the effects of gravity.

DAREDEVILS

Using a drinking vessel made from pressed glass is an opportunity to create a more long-lasting vacuum. Half fill a shot glass with liquor. Set it alight, blow it out, then immediately place an outstretched palm onto the top of the glass. Glass and hand will be held fast but can be separated by a gentle pull, followed by a discernible popping sound.

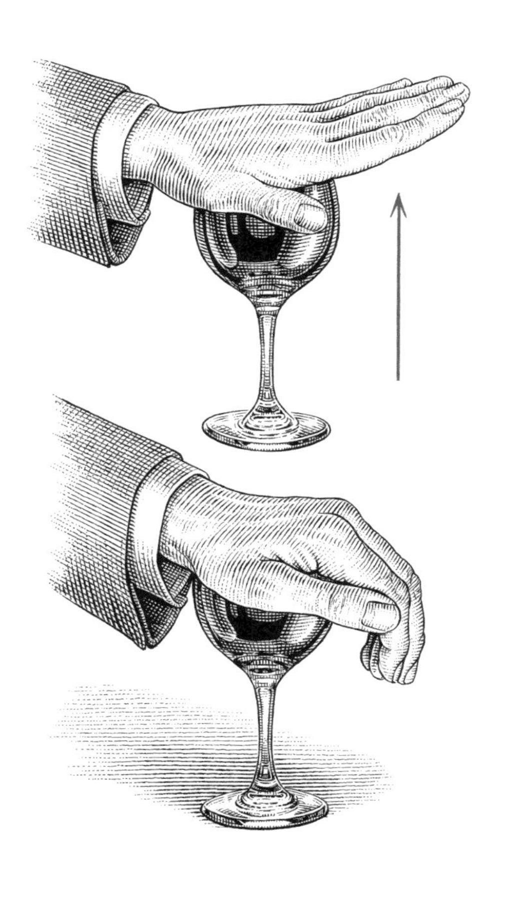

IN WHICH THE HOST
PERFORMS A LEVITATION

IN WHICH THE HOST HANGS,
DRAWS AND QUARTERS A ROGUE DESSERT

THE GUILLOTINED PUDDING

EQUIPMENT

Ripe orchard pear, fine string or thread,
a taper or long match, a sharp knife or two

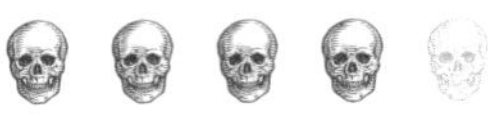

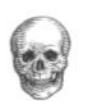

WHEN LEAVES FALL in autumn, so do pears: sometimes all at once, if you allow them to ripen on the tree. If a falling pear happened to land on sharpened blades cannily arranged, the gourmand's classic combination of cheese and pear would be more expediently prepared.

The best kind of pear to eat, and to drop from a considerable height, is a large, plump one. Pears bought in shops are acceptable, though the best are not usually available for purchase – it is all too likely that something narrow, like a Conference pear, will be on offer. Find out which of your friends has a tree of Doyenné du Comice or Williams and invite them to dinner.

Select a perfectly ripe specimen of good overall shape. Tie a length of fine string or thread to the stalk, knotting it tightly so that it doesn't slip off. A perfectly ripe pear should give a little when pressed around the stalk. Too ripe – with the slightest hint of mushiness – and the stalk will detach along with the top of the pear.

Hang the pear as high as you can, over a table or the floor. When no one is looking, wet the bottom of the pear in a glass of water and watch where the drips fall. Mark the spot discreetly – this is the point over which the knife should be held. While one guest holds a knife – blade up – another lights a taper (though a match will do) and sets fire to the string. Wait patiently with a plate and gravity will do the rest.

DAREDEVILS

Prefer a cruciform of knives, as illustrated here.
A very ripe pear will be juicier, messier and tastier.

TRIVET of TRIUMPH

EQUIPMENT

Pudding dish or serving bowl, large dinner plate, three clean forks, silver napkin ring

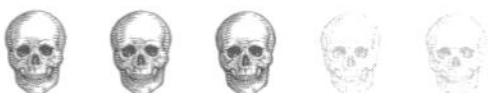
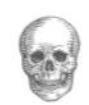

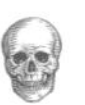

TRADITIONAL SCHOOL PUDDINGS, recalled with fondness or horror, are an indelible experience. Any devotee of custard would feel even more blessed when given chocolate sponge floating in chocolate custard. They might even appreciate the finer points of rice pudding or semolina, though this kind of connoisseur will find themselves in a tiny minority.

An essential point about nursery food is that it must be eaten hot, before a skin forms and lumps become lumpier. And a hot dish needs a trivet, like the elegant example shown here. It is a picture of simplicity, but how simple is it to assemble?

The hot pudding dish is held by an impatient cook while the host grabs a fork from neighbours on either side and adds their own fork, plus a napkin ring, to demonstrate how to create a fully functioning stand. It's really quite easy: balance three forks pushed through a napkin ring, equally spaced, the prongs stopping at the lip of a plate that in turn holds the bowl.

The instant it has been made this handsome trivet should then be disassembled at speed. The host takes the pudding from the plate and holds it impatiently while a guest grabs the forks, napkin ring and plate, attempting to disassemble and assemble as quickly as possible, before passing along all the elements of the trivet to their neighbour and holding the pudding themselves. Each competitor should be put under undue pressure, with general jeering and distracting commentary.

DAREDEVILS

Will attempt this with hot soup.

IN WHICH THE HOST
FASHIONS AN IMPROMPTU TRIPOD

IN WHICH THE HOST
REDRESSES THE TABLE

TABLE TURN TOSS

EQUIPMENT

Fully laid table, slippery tablecloth (hemmed on the long sides only)

 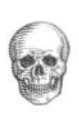 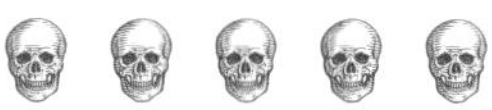  

SOME AFTER-DINNER EXPERIMENTS can be approached as performance rather than group activity. Circumstances might dictate, however, that a host is in possession of a detested set of china, the destruction of which would be a happy outcome. In this case, anybody can have a go.

This performance, or demonstration, is essentially a variety-show trick of uncertain vintage. It may go back as far as Sir Isaac Newton, who first noted the effects of inertia. An object, he observed, will always stay put. Unless there are outside forces at work – for instance when the stationary object is pushed down a slide, or when a cat knocks it off a high shelf. Objects resting on a table are inclined to remain there, even if the surface on which they sit is moved. If this surface happens to be a tablecloth, friction would be to blame for any precipitation.

For a clean sweep, friction needs to be minimal. A silky, slippery tablecloth is to be recommended. For an oblong table, it is important that the two short sides of the cloth be unhemmed. (It would not be unreasonable to have this cloth specially made.) The tabletop must be perfectly smooth, with no adjustable panels breaking the line, and standing level on the floor.

Ask guests to stand at a respectful distance and then firmly grasp the cloth (on a short, unhemmed side) with both hands. Whip – literally – the cloth out from under the place settings, pulling it straight down towards the floor, with great alacrity. Remember, practice makes perfect.

DAREDEVILS

Put the tablecloth back on, by following the instructions in reverse. Similar levels of speed and dexterity are required.

EQUIPMENT

Cylindrical, shouldered glass bottle, a straw (ideally paper)

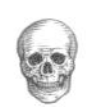

A SIMPLE ENGINEERING EXERCISE that can be tricky, as well as sticky, if rushed into. Though it works just as well with a beer or cider bottle, it is a good pastime for drinkers of alternative beverages, such as lemonade.

Sturdy, stripy paper straws are not as readily available as bendy plastic straws, though they really should be. Therefore, snap them up whenever you see them (for instance, in the party section of a department store), not least for their comparable efficiency and superior looks. For this kind of lever action, a paper straw comes into its own. A bendy straw is not impossible, if it is the only thing to hand.

Fold a straw about 5 centimetres from the end (the bend in a plastic straw is not useful here; use the other end). Gingerly lower the folded straw into an empty or near-empty bottle of medium size. As the straw unfolds, the short end will get stuck in the bottle neck, forming a simple lever. As long as the paper straw remains dry, it will hold fast. (A plastic straw will soon give up the struggle to support any kind of weight, and a bendy straw, pushed in at the bendy end, will simply unbend.)

The weight of the glass bottle is distributed along three points of the lever. The longer end of the straw, rising above the top of the bottle when pulled, is the effort arm. The short end of the straw forms a resistance and acts as the load arm. The bend is the third crucial part: it is the fulcrum, the fixed part that transmits the effort and the load.

DAREDEVILS

Know that a lightweight but sturdy lever can lift a heavy object, including a bottle that is more full.

IN WHICH THE HOST CARRIES
THE BOTTLE OFF WITH APLOMB

THE

Sommelier

and the

SABRE

EQUIPMENT

Sparkling wine in a thick bottle, a long, flat knife

 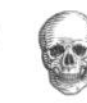 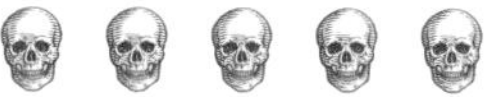 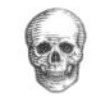 

A SABRE, HELD ALOFT while charging, was as important to a nineteenth-century hussar as his tight trousers, moustache and braided tunic. While this military style was adopted all over Europe, Napoleon's hussars may have had the most *joie de vivre*, slicing open champagne bottles at every opportunity. Today's hosts are less likely to have a sabre on hand than a sturdy chef's knife – the perfect aid to occasional sabrage.

In choosing equipment, it is important to remember that it is not a *weapon* that is useful here, but a blunt-edged instrument. While a spatula lacks the romance of a sword, it would be more effective; a sabre has only one sharp edge and that edge is redundant. The experienced sommelier knows that a combination of exterior impact with interior pressure allows the top of the bottle to be sliced off with ease.

Chill the champagne, keeping the neck the coldest. Step outdoors, keeping guests safely behind you. Dry the bottle, removing foil and cage, while holding the cork with your thumb. Next, find the seam that runs the length of the bottle and hold it out in front of you at an angle of between 20 and 30 degrees. (The junction of seam and neck is the bottle's weakest point.) Turn your knife over so the blunt end is facing the top of the bottle. With one deft motion, sweep the knife swiftly along the seam and carry straight on through the neck. Do not lose your nerve halfway – follow through. Let the wine flow a little bit to disgorge any shards of glass before pouring.

DAREDEVILS

Risk drinking from the first glass
(after carefully checking it for small shards)

IN WHICH THE HOST
CUTS TO THE CHASE

IN WHICH THE HOST
PERFORMS A MINIATURE HULA

Spinning HALOES

EQUIPMENT

Lightweight napkin ring, bottle of good ordinary claret

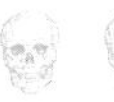

CAN A DINNER party be taken seriously when there are paper napkins involved? Not unless it's a picnic. Before the age of convenience, linen napkins encircled by napkin rings were also held in contempt, since they implied an economy of laundry. Engraving of initials or worse, numbers, betrayed the fact that napkins were used more than once, each assigned to different family members, servants or lodgers in a boarding house. 'I have never paid attention to the shibboleths of the fashionable world,' observed Osbert Sitwell in the second half of the twentieth century, 'but I remember being brought up to regard napkin-rings as very vulgar.'

While a starched and beautifully folded napkin would be more usual at a formal dinner, the fashionable world of today is unembarrassed by napkin rings. Once a napkin is unsheathed, these bourgeois devices can be requisitioned as rings for finger spinning after dinner. When King George V first encountered a napkin ring, he exclaimed, 'What's this? It's far too big for my finger.' In fact, it was the perfect size.

Stand by the centre of the dining room table and place your finger in the middle of a napkin ring. Start spinning, quickly and evenly, and lift the ring into the air. The centrifugal force allows you to keep the halo-like ring airborne, pushing the ring away from your finger and resisting the force of gravity as the direction of movement pulls the ring around. See how high you can propel the ring through this spinning action, attempting to drop it over the neck of a nearby bottle.

DAREDEVILS

Place a bottle of wine at each end of the table. Spinning a ring with each hand, collar the two bottles at the same time, without knocking anything over.

THE BENIGN FLAME

EQUIPMENT

100 to 140 proof alcohol, paper banknote, tongs (optional)

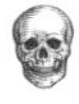

BURNING MONEY FOR pleasure raises various questions: Is it legal? (Yes in the UK, no in the US.) And how does it affect the wealth of the nation? (If you set fire to regular paper currency and it really does burn to cinders, then its removal from circulation will increase the value of all remaining currency by a minuscule amount.)

Anybody who has found banknotes in their pocket after doing the laundry will know that paper money isn't made of paper. All American and some British bills are made from cotton fibres, blended with linen. They are absorbent, which is part of the reason they will survive mock immolation. Generally this trick is performed with the aid of surgical spirit. It's efficient, being at least 50% alcohol by volume, but it's not generally on hand in a dinner situation, so a drinkable spirit would be better. Whisky, vodka and gin are usually bottled at 40% ABV, but something between 50 and 70% is ideal: a strong grappa, tequila or rum. Grain alcohol is TOO strong, ditto absinthe.

Soak a bill in a tumbler mixed half and half with water and spirits. (Add salt for a more dramatic effect: yellow flames instead of blue.) Lift the money out with a pair of ice tongs and set it alight. Flames will run all over the note, like a Christmas pudding. When they burn themselves out the money will not even be hot. The cellulose fibres of the paper currency absorb water. Alcohol burns at a low temperature, so the flame will go out before the water evaporates. A note on 'plastic' or polymer bills: according to the Bank of England, they are 'cleaner, safer and stronger'. They are of no use here.

DAREDEVILS

Don't bother with the tongs.

IN WHICH THE HOST HAS MONEY TO BURN

IN WHICH THE HOST
DISPENSES WITH THE CHAIRS

LA

TABLE VIVANTE

EQUIPMENT

Four guests, four dining chairs

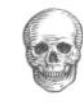    

ACROBATICS INVOLVING CUTLERY and food are all very well but what of personal physical prowess? Even after hours of eating (and drinking), a feat of physical strength can add mirth and merriment. Dinner party levitation is one route, with unlikely results, depending on the power of concentration. Substitute corporeal flexibility for magical thinking and you have the Living Table.

Four volunteers are required. To test their fitness, see if they can do the crab pose, even if they do not practise yoga. This involves lifting the chest and hips off the ground so that only the palms and feet remain on the floor – it will give some idea of what is required.

Arrange four chairs as though pulled in under a square table. The chairs need to be close enough that when one person, with feet on the ground, lies down sideways across a chair, the adjacent chair will be able to support that person's head and shoulders. Next, have all four guests sit down so that their bodies are at right angles to the chair-backs. One by one, each person lies down, resting their head and shoulders on the next person's legs.

A bystander removes each chair, one by one, and the human table does not collapse: combined and balanced forces keep your friends intertwined and aloft. The chairs can be put back at some point; alternatively, an end can be hastened by asking the members of the table to scuttle sideways, like a crab, or rotate around, like an unwieldy jigsaw puzzle piece.

DAREDEVILS

Mix a round of cocktails and place the tray on the human table, entreating guests to partake.

EGG on a Tightrope

EQUIPMENT

Hard-boiled egg, cork (not synthetic), two sturdy forks, bottle of wine

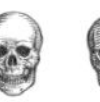

PICTURE THE CIRCUS. High above you, a tightrope walker in tights and tutu traverses the big top, with or without the security of a net far below. A pole is used for balance, carried more or less horizontally. Now imagine this performer as an egg, upright and poised. In terms of equilibrium, there is little difference between the two.

Instead of a pole, two forks and a cork come to the aid of this performing egg. Take two identical forks of equal weight and insert their prongs into either side of an upright cork, its narrower end pointing downwards. (When in place, the forks and cork will have a similar silhouette to a winged corkscrew.) With a penknife, make a slight indentation in the lower end of the cork, so that it makes a comfortable fit when the ensemble is placed on top of the base of the egg.

Place the egg (pointy side down) and its cork and forks on the rim of the bottle. With a few small adjustments and a steady hand the whole thing will balance upright. The centre of gravity for the stunt egg is the bottle rim, where the weight of the whole assemblage is evenly distributed. Resting the egg in the mouth of the bottle is not the point.

IN WHICH THE HOST
HANGS AN EGG IN THE BALANCE

IN WHICH THE HOST
MAKES SPORT OF GREEDY GUESTS

Ordeal of the SUGAR CUBE

EQUIPMENT

Sugar cube, solid high-backed chair

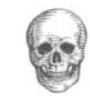
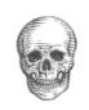

FOR THOSE OF us who inherit furniture that we don't particularly want but feel too guilty to throw away, this trick could give new purpose to granny's mahogany chairs. Alternatively, a simple utilitarian seat, solidly constructed, will work just as well, as long as its planes are parallel.

It sounds an easy proposition, to grab a sugar cube between your teeth. But anyone who remembers the chocolate game – in which you may eat as much chocolate as you like until the next person rolls a double, but first you must don hat, scarf and gloves, and you have to use a knife and fork – will be familiar with the feeling of being hopelessly tempted as the dice are rolled and the chocolate taken away. So near and yet so far. It's a kind of hell, reminiscent of the punishment of Tantalus in the underworld: because of his outrageous hubris, he was doomed to spend eternity teased with the food and drink he craved just out of reach.

In this case, the object of desire is a sugar cube. Lay a straight, high-backed chair on the floor horizontally so that the back forms a kind of platform. Place a sugar cube on the top rail and ask a volunteer to climb onto it, shins resting against the chair legs, while grasping each side of the chair-back. As the volunteer inches towards the goal, the chair falls forward and the sugar cube rolls away. The dinner guest lurches dramatically forward, possibly landing on their head. Be careful with teeth.

CANDLE QUADRILLE

EQUIPMENT

Candles, heavy candlesticks, matches

BALANCING ON ONE knee is an unusual thing to do, but with a little practice it can seem quite natural and even zen-like. The introduction of fire challenges this happy equilibrium. Consider the surface of the ground: a carpet is far preferable to wood or stone, though perhaps it should be covered with newspaper. Linoleum never looked so enticing.

Both participants kneel on the floor opposite each other and lift their respective right knees behind them, grasping their right ankles with their right hands. A third person places a candle in each of their left hands and lights one of them. Candles in candlesticks (or ideally candelabra) will be heavier for holding aloft, while pushing the wicks further away – this makes for more exciting spectating. One of the candles is lit. The object is for the participants to approach each other and then for one to light the other's candle without either of them letting their right knee touch the ground.

DAREDEVILS

Add another step: once both candles are lit, the third person blows one out. The kneelers must relight the candle without touching the flame to the wick. (The trick is to hold the flame to the smoke – the flame will ignite the candle vapours and light the other wick. Note that paraffin candles must be used: this does not work with beeswax candles.)

IN WHICH THE HOST
PASSES ON THE TORCH

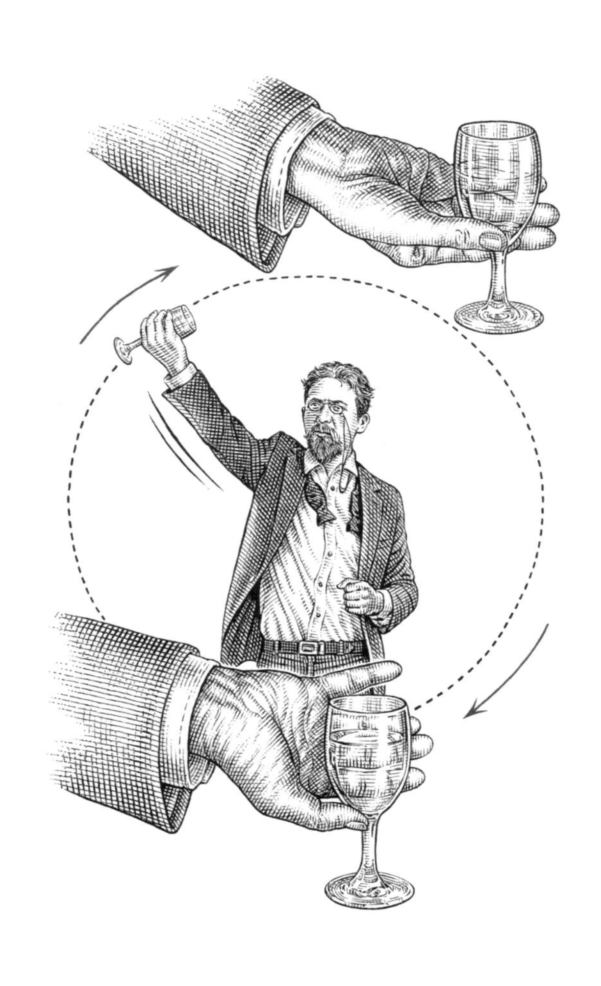

IN WHICH THE HOST
DESCRIBES A VINOUS CIRCUIT

WINE REVOLUTION

EQUIPMENT

Wine glass, wine (or water)

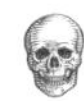

A POPULAR DINNER PARTY guest knows that a little Latin goes a long way. An intimacy with epic poetry is less important than a few common expressions, as well as some indispensable verbs, viz. *petere* (to seek) and *fugare* (to flee).

The key to success with this trick is in holding the glass correctly. Grasp a semi-full wine glass with thumb and fingers inverted so that they grip the bowl, not the stem. Your little finger will be at the top, forefinger at the bottom, wrist facing away. Revolve your fully extended arm, like a bowler in cricket, maintaining a steady speed.

The glass begins its journey upright, though halfway through a full revolution it will spend some time upside down. Yet only if you falter will the liquid fall, since it is inclined to travel in the same direction as the glass when the glass is in motion. This is the first law of motion according to Sir Isaac Newton, also known as the law of inertia. In other words, unless acted upon, a stationary object will remain stationary and a moving object will remain in motion, because of balanced forces.

To avoid explanations *ad absurdum*, just think of the moon orbiting the earth. It is pulled towards the earth's centre by gravity, a centri*petal* force (for those conversant in Latin, a force seeking the centre) but it is kept away by a centri*fugal* force (a force fleeing the centre). These are balanced forces and they keep the wine in the glass.

The trickiest part is starting and finishing, while gaining or reducing speed. There is no harm in a little practice ahead of time. Remember: *ars longa, vita brevis*.

DAREDEVILS

Perform this with red wine.

TEACHING A BOTTLE TO SUCK EGGS

EQUIPMENT

Carafe or wide-mouthed bottle,
hard-boiled egg, scrap paper, matches

A USEFUL IMAGE IN pondering the physics of eggs and bottles is of a person being sucked out of an airplane in mid-flight. Airplane cabins are pressurized because the air pressure outside at altitude is too low for human well-being. When someone opens an airplane door in a disaster movie, the higher pressure in the cabin pushes them out, leaving no time to reconsider.

A boiled egg's progress through the atmosphere is more sedate, but it needs to be thoughtfully prepared (ideally before guests arrive) for its journey, its smooth shape unimpaired by rushed boiling or clumsy peeling. Set a timer for 13 minutes instead of the usual 6 and turn down the heat. Only after this time take the hot pan to the sink and run cold water over the egg until it is significantly cooler. Add ice if guests are waiting impatiently.

The egg should perch comfortably on the mouth of the bottle. Gently crack, roll and peel it. A bit of moisture will speed its progress – popping the egg into your mouth when no one is looking will accomplish this expediently. Light a strip of paper and drop it, along with the match, into the glass vessel. Replace the egg immediately and marvel as the bottle seems to ingest the egg of its own volition.

The heat of the fire briefly increases the pressure inside the bottle. When the flame goes out, the air pressure inside the bottle decreases; the greater outside pressure pushes the egg in, restoring equilibrium.

DAREDEVILS

Tip the vessel upside down and blow into it like
a trumpet, risking egg on the face when it pops out again.

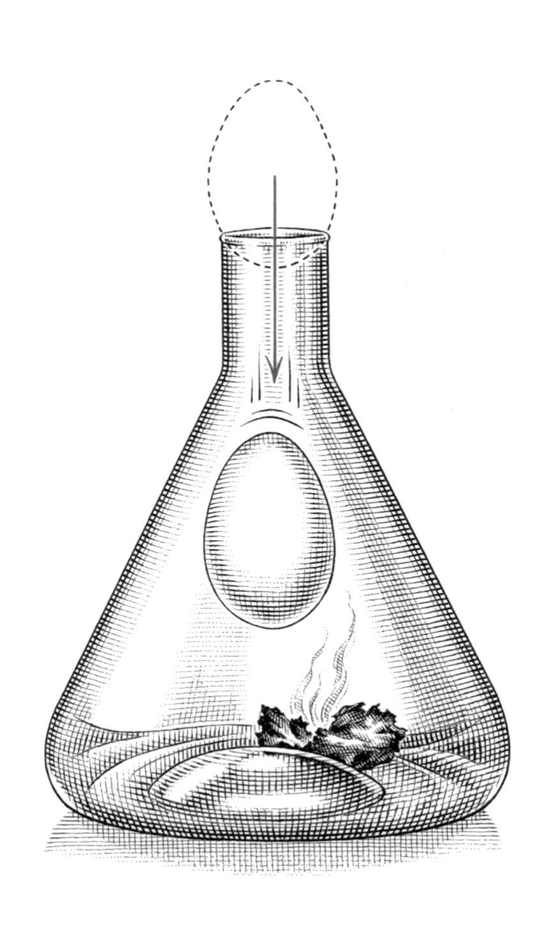

IN WHICH THE HOST
PERSUADES AN EGG INTO A BOTTLE

IN WHICH THE HOST CONTRIVES
A GRAVITY-DEFYING SPECTACLE

EQUILIBRIUM of a TOOTHPICK

EQUIPMENT

Wine glass, two lightweight forks, toothpick, matches

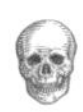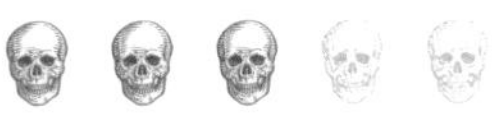

ON THE SUBJECT of table etiquette, forks easily attract the most ridicule. Before the Victorians complicated matters by designing dozens of slightly different shapes for any contingency, forks were barely used at all. They were practical in cooking but at the table what could a fork do that fingers couldn't? This is a question Americans still ask when watching Europeans of a certain vintage battling a hamburger with knife and fork.

The balancing trick described here requires two identical forks of equal weight. It is best suited to people who would never think of using silver or silver plate at the table. Even superior stainless steel is not appropriate here. Since the tines of the two forks need to be interlocked, their long handles floating in the air, the only way to satisfactorily accomplish this is with cheap, lightweight forks.

When you've woven the prongs together, balance the forks evenly on your finger where they intersect. The point at which equilibrium is achieved is the place where you need to insert – or rather force through – a toothpick. (It should emerge at the other side.) Balance the long end of the toothpick over one part of the circumference of a glass, until you find its pivot point. Once the centre of gravity (the point at which weight is evenly distributed) has been established, the whole contrivance will balance quite happily.

DAREDEVILS

Set fire to the toothpick inside the glass. The flame will go out as it touches the glass and the forks will appear to float.

THE SCOTCH HOP

EQUIPMENT

Two straight-sided crystal tumblers, single or blended whisky, a non-porous card

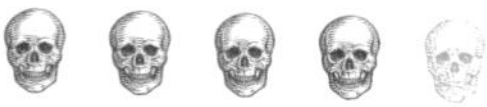

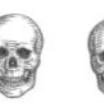

JUST AS THE weight and quality of a crystal tumbler needs to be taken into account when drinking whisky – whether neat, on the rocks or in an Old Fashioned – so too does the pair of glasses needed for this after-dinner experiment. Sipping and experimenting can take place in parallel, since after the initial bout of concentration and set-up the whisky looks after itself.

Fill a cut crystal tumbler to the brim with whisky and set it down on the table. Fill a second, identical glass to the brim with water and completely cover it with something slim and non-porous, such as a laminated card or postcard. (Note: a thick invitation will create too wide a gap between the glass rims.) Invert the water glass and place it directly over the first, fitting one rim exactly above the other. Carefully, gently, pull the card over slightly so that the liquids are just exposed to each other.

Water is heavier than whisky, so it will flow straight down, creating a pleasing visual effect. After the water has finished dropping into the whisky, causing the latter to move upwards, the contents of the two glasses should be completely transposed – a rough mixture of whisky and water, while an acceptable drink, is not the desired result. Gently replace the card so that it completely covers the mouths of both glasses and remove the whisky glass. Restore the tumbler to its upright position and enjoy.

DAREDEVILS

Experiment with rare single malts.

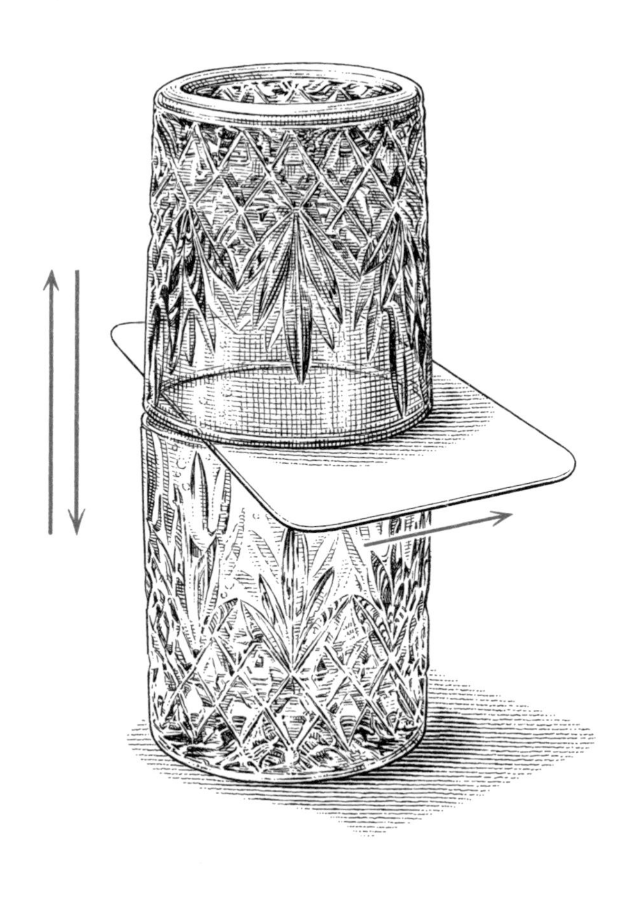

IN WHICH THE HOST
TURNS WATER INTO WHISKY

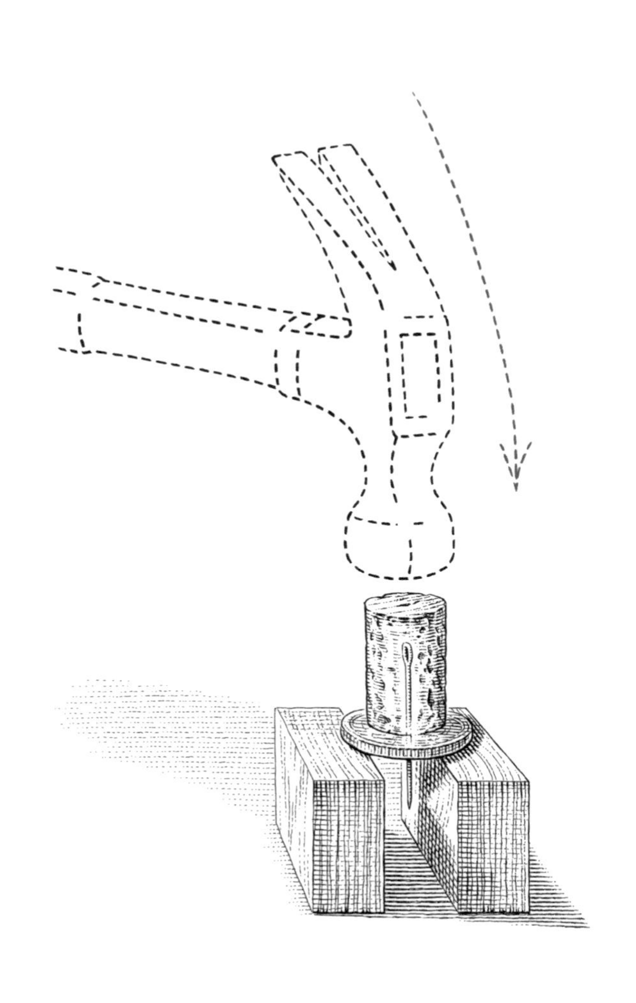

IN WHICH THE HOST NAILS IT

PIN MONEY

EQUIPMENT

Sturdy hammer, needle (100% steel), cork and supports, alloy coin (0% steel)

HAMMERS ARE MAINLY associated with daytime activities, swept away in the great tidy-up before guests arrive. Segue seamlessly from dinner to after-dinner hammering by serving something that requires a hammer at the dinner table, for instance crab or lobster. Although being forearmed is as good as being forewarned, advance notice of this activity could be counterproductive.

The challenge for your guests is to pass a needle through a coin. If they know their densities, this is perhaps simple enough – the solid steel needle is denser by far than the alloy coin. But how to keep the needle from bending or even snapping when struck a blow with a hammer?

If there are children or former children in the house, a pair of wooden building blocks would be ideal as supports. Otherwise, two unloved books of similar size would do. Place these side by side with a gap between, like bluffs on either side of a canyon. Across this canyon, place an American penny, nickel or dime. A British penny or tuppence is redundant here, though a 20p piece is acceptable.

Find a needle in advance – choose a long one so that when it is pushed into a cork the end with the eye stands slightly proud of one end of the cork (half of the eye should be visible). The pointed tip should only just protrude from the cork's other end.

Place the needled cork directly on top of the coin. Hammer the needle with firm but gentle thrusts. 'Gentle' is the key word here. The cork will flatten slightly but will keep the needle straight as it pierces the coin and proceeds through it.

DAREDEVILS

Match their strength to that of an old British one pound coin.

STRONGMAN SWIZZLE STICKS

EQUIPMENT

Three wide, conical champagne flutes, two cocktail sticks, a tablecloth for a soft landing

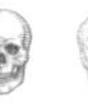

A WELL-STOCKED BAR IS the sign of a well-oiled host, one who realizes that the high moments of life are to be toasted with the aid of under-used drinking accoutrements and récherché cocktail ingredients. Glass cocktail stirrers or swizzle sticks could be put to further use in assisting cocktail-hour acrobatics. For the more casual host, chopsticks cut in half are a fine substitute, and for al fresco occasions twigs will add a different dimension.

Starting with a wider sort of conical champagne flute standing upright, place a glass cocktail stirrer, or stick, inside. Take an identical flute and balance it sideways, using the stick to 'hook' it from the interior and hold it so that the floating glass stays in place without the need of steadying hands. With one glass thus hoisted repeat this step to add another to the other side.

This is an experiment that demonstrates more than balance: it shows a cantilever in action. In other words, a load-carrying beam, supported only at one end. One stick end remains in a fixed place at the bottom of the glass, while the other is free to hold an unlikely weight. Think of a balcony, the Brooklyn Bridge, a floating staircase.

NEVER TRY THIS WITH PLASTIC CUPS – YOU'RE ONLY CHEATING YOURSELF

DAREDEVILS

Continue building with further swizzle sticks and additional glasses.

IN WHICH THE HOST ENGAGES
IN A SPOT OF CONSTRUCTION

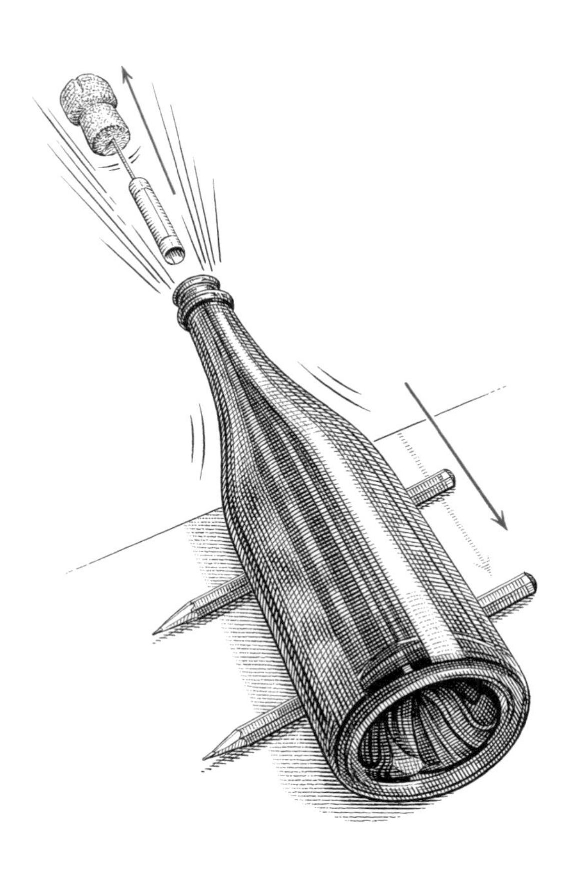

IN WHICH THE HOST
DISPOSES OF THE CORK

BATTERY of BUBBLES

EQUIPMENT

Champagne bottle (empty), cork, citric/tartaric acid or cream of tartar, bicarbonate of soda, a spare playing card, string, tape, kitchen paper, pin, two pencils (optional)

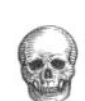

IN THE NOVELS of P.G. Wodehouse idle members of the Drones Club would have pulled off a prank like this between lunch courses. However, without the ministrations of a gentleman's gentleman, the tabletop cannon salute does involve some minimal forethought. This is mainly in the obtaining of citric or tartaric acid, a vital ingredient in cordial-making and still purchased from a chemist. In the absence of this, cream of tartar, along with bicarbonate of soda, may be readily found in any well-stocked larder.

Take an empty champagne bottle and half fill it with water. Dissolve into it one tablespoon of bicarbonate of soda. Place a teaspoon of tartaric acid (or a double quantity of cream of tartar) on a playing card and roll it up, using kitchen paper as a plug. Attach some string to this cartridge-like contrivance, then tape it all together. The other end of the string needs to be tied around a drawing pin, which is pushed into the base of the cork. Make sure that when the bottle is standing upright the cartridge dangles over the water without touching it, the contents plugged in.

Only place the bottle on its side when you are ready. To get the full *coup de canon*, place the recumbent bottle across two parallel pencils, like runners. When the water penetrates the cartridge, it dissolves the tartaric acid and the resulting reaction fills the bottle with carbonic acid gas, pushing the cork out in a great shot. Like a discharging cannon, the bottle will roll backwards slightly on its runners.

DAREDEVILS

Use double quantities of powder.

CORK
and
FORK
PIN SPIN

EQUIPMENT

Dinner plate, three corks, bottle of wine, four good forks, stout needle

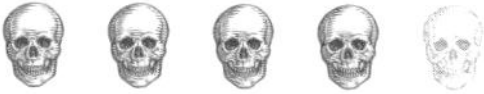

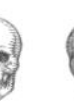

THIS EXPERIMENT IN angular momentum has an outlandish appearance that places it firmly in the past. In the Age of Enlightenment, when Sir Isaac Newton's laws of physics were fresh and Josiah Wedgwood was introducing good china to the middle-classes, the messy consequences of dining room experiments were like so many angels dancing on the head of a pin. These days, broken porcelain can lead to broken friendships. Less bold souls may wish to try this at a picnic with melamine plates.

Like spinning plates that lose balance and fall off their pivots when they slow, or a unicycle that fails to hold the pedaller aloft once the wheel stops moving, the cork and fork pin spin relies on the speed of rotation, or angular momentum. Carefully push a needle, eye first, into a cork that is already stuck into a wine bottle. Find two other corks and cut them each in two, lengthways. Push the prongs of four forks into the flat ends of the four cork sections. It is important that the fork prongs be sharp. Place the forked corks at even intervals around the plate, so that the forks are dangling at a slightly acute angle.

Put the plate on top of the needle and start it rotating, finding the pivot point as you do so. The reason for using a needle is to introduce minimum friction and thus increase spinning velocity. You may find that the needle's positive qualities are outweighed by its negatives, but just bear this in mind: the faster the spin, the greater the stability. And please, no plastic cutlery.

DAREDEVILS

Send steak knives and porcelain on a heady fair ride.

IN WHICH THE HOST DEVISES
A POSTPRANDIAL WHIRLIGIG

Published by
Laurence King Publishing
361–373 City Road
London EC1V 1LR
United Kingdom
T +44 20 7841 6900
enquiries@laurenceking.com
www.laurenceking.com

A catalogue record for this book is available from the British Library.

ISBN: 978 1 78627 617 9

Printed in China

Laurence King Publishing is committed to ethical and sustainable production.
We are proud participants in The Book Chain Project ®
bookchainproject.com